Australian GEOGRAPHIC

Australia's Amazing
Sharks

Australia's Amazing
Sharks

First published in 2020
Australian Geographic
52-54 Turner Street
Redfern NSW 2016
editorial@ausgeo.com.au
australiangeographic.com.au

Funds from the sale of this book go to support the Australian Geographic Society, a not-for-profit organisation dedicated to sponsoring conservation and scientific projects, as well as adventures and expeditions.

Shark illustrations: © R.Swainston/ANIMA.fish; Marje Crosby-Fairal/Australian Geographic and Kevin Stead/Australian Geographic
Creative director: Mike Ellott
Senior designer: Harmony Southern
Editor: Rebecca Cotton and Lauren Smith
Sub-editor: Peter Tuskan
Editorial intern: Jasper Steel
Print production: Katrina O'Brien

Australian Geographic Managing Director:
Jo Runciman
Australian Geographic Editor-in-chief:
Chrissie Goldrick

Contents

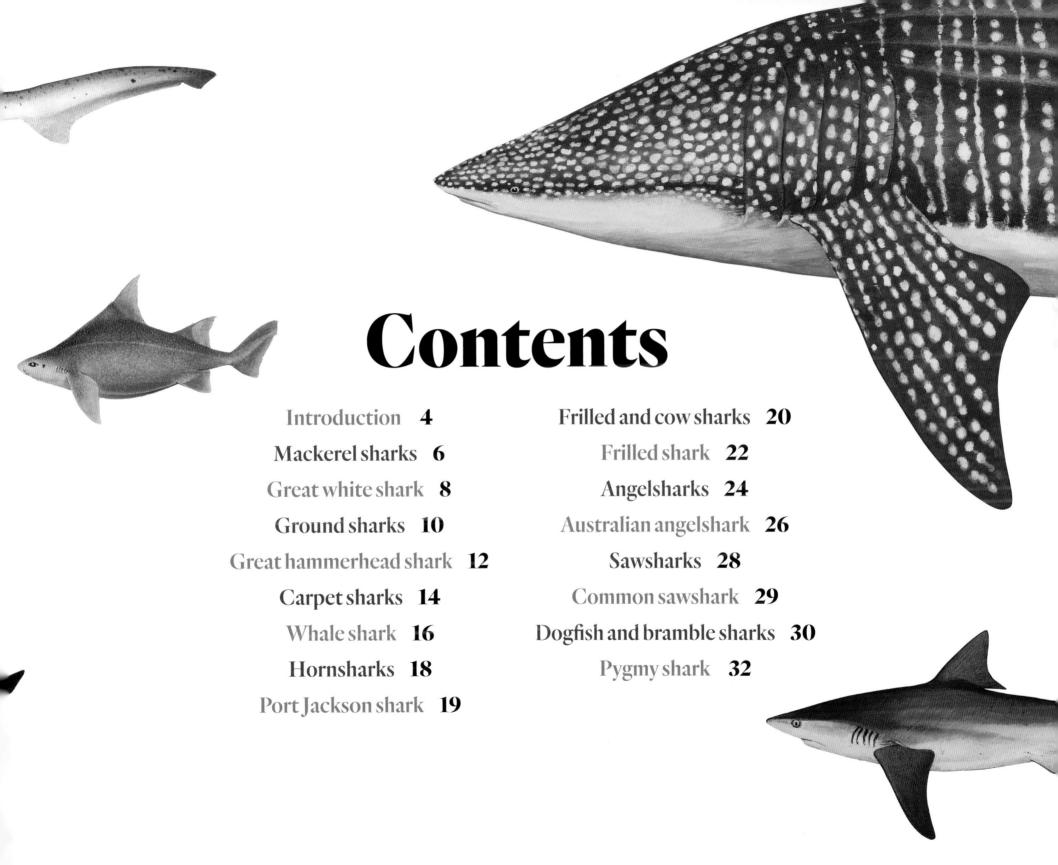

Introduction:

Sharks

MORE THAN 400 species of sharks are found worldwide today. About 170 of them inhabit Australian seas, from the world's largest, the whale shark, to one of the smallest, the pygmy shark, and of course, the equally fascinating and fearsome great white. The Coral Sea, off the Queensland coast, is a particular hotspot of shark diversity, with more than 50 species found there.

Perfectly adapted to a life under water, sharks have patrolled the oceans for more than 400 million years. It takes a very well designed animal to survive such a long time and sharks have developed some amazing abilities to help them become top ocean predators.

Although sharks can't see as far as they can smell or hear, they still have excellent sight in the water. Most sharks can see better in dim light than their prey. They also have

an incredible sense of smell, particularly for blood, which they can smell from a long way away. Sharks can smell the equivalent of one teaspoonful of blood in an Olympic-sized swimming pool!

Sharks also have one super sense that humans don't have: they can pick up electrical signals. This is called

electroreception. All animals, even tiny ones, give off electrical signals when they move. Sharks pick up these signals using tiny receptors in their snouts. They can also sense movement or vibrations in the water, using a special system of vessels that run underneath their skin. That's why sharks are sometimes attracted to animals (including humans and dogs)that are kicking or splashing in the water.

Shark populations around the world have plummeted as the human population has skyrocketed. Largely due to over-fishing, more than 200 shark species are currently listed as threatened, and 50 as vulnerable. Only three shark species (the great white, whale and basking sharks) are protected internationally.

THAT'S AMAZING

Sharks have up to seven rows of teeth. Every time a shark loses or wears down a tooth, the one behind moves forward to take its place. Some sharks will lose up to 35,000 teeth in a lifetime.

Mackerel Sharks

This group of powerful, fast-swimming sharks can be found in oceans all over the world, and includes some of the most famous species. Mackerel sharks can be identified by a number of features, including having five gills and a mouth that extends back past their eyes.

❹ Basking shark
Cetorhinus maximus
Length: 10m **Status:** Vulnerable

This is the second-largest living shark. It feeds on tiny fish by sucking water in through its mouth and catching the fish in its filters. Its big nose has given rise to one of its other names – the elephant shark.

❸ Megamouth shark
Megachasma pelagios
Length 5.5m **Status:** Least concern

When first encountered in 1976 this creature was dubbed the 'megamouth' due to its gaping mouth and huge jaw, both of which are much larger than the shark's abdomen. This bulbous-shaped head means it isn't the fastest swimmer. Believed to be active mainly during the day, the megamouth regularly alternates between the shallow and deep waters around Japan and China, but has also been observed travelling from the Atlantic Ocean right through to the Pacific and Indian Oceans.

❶ Goblin shark
Mitsukurina owstoni
Length: 3.9m **Status:** Least concern

With an extremely long snout, long jaws and a pinkish body, the goblin shark is very strange-looking. It lives in the very deep parts of the ocean, where there's no sunlight. That means it can't see its prey – instead, it picks up on their electric signals. When it gets close, it sucks its prey in close to its mouth and then gobbles it up.

❷ Crocodile shark
Pseudocarcharias kamoharai
Length: 1.18m **Status:** Least concern

The crocodile shark is found in tropical and subtropical waters of all oceans. Its common name refers to its sharp teeth and habit of snapping vigorously if captured. It feeds on fish, molluscs, crustaceans and squid, staying deep in the ocean by day and surfacing at night.

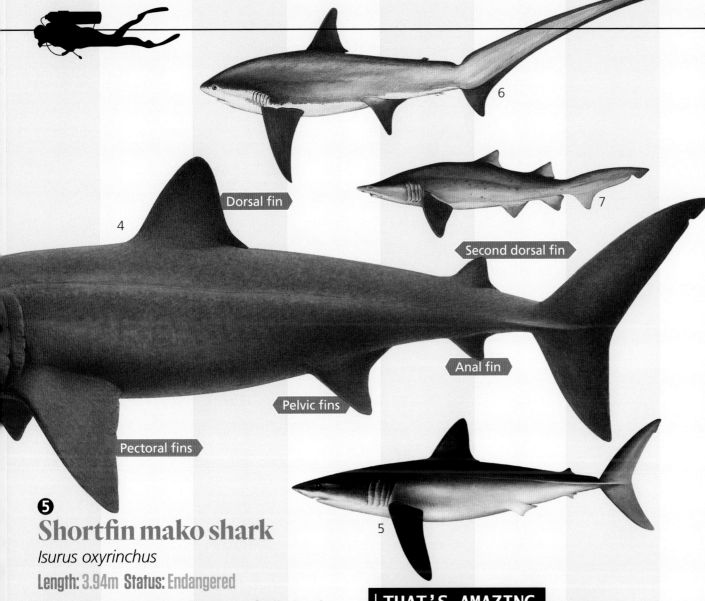

Dorsal fin

4

Second dorsal fin

Anal fin

Pelvic fins

Pectoral fins

6

7

5

❻ Thresher shark

Alopias vulpinus

Length: 5.7m **Status:** Vulnerable

Immediately recognisable by its long tail – which it uses to herd and stun squid and schools of fish – this shark can leap up to 6m out of the water. It is widespread and common in tropical and temperate waters worldwide. In Australia, its range stretches south from Broome, Western Australia, all the way to Brisbane, Queensland, including Tasmania. This is a large, powerful shark, though it is not known for being aggressive towards humans.

❼ Grey nurse shark

Carcharias taurus

Length: 3.18m **Status:** Vulnerable

Protected in New South Wales and Queensland, the grey nurse looks fearsome thanks to its exposed teeth, but is not considered dangerous. It's found in all Australian waters except around Tasmania. It eats fish, rays, squid and crustaceans, and is often seen close to the sea floor.

❺ Shortfin mako shark

Isurus oxyrinchus

Length: 3.94m **Status:** Endangered

Thought to be the fastest of all sharks, the large, torpedo-shaped shortfin mako is capable of bursts of speed up to 74km/h. A popular target for big-game fishing, this shark can leap up to 6m out of the water.

THAT'S AMAZING

While they're still growing inside their mother, baby grey nurse sharks eat each other until there is only one left.

Great white shark

Carcharodon carcharias

Status: Vulnerable

Length: 6m

PROTECTED THROUGHOUT Australian waters, the largest carnivorous shark in the world favours cool, shallow, temperate seas, and is most commonly found in southern Australian waters from Exmouth in Western Australia, to southern Queensland. It feeds on fish and marine mammals such as seals.

The great white is pelagic, which means it lives deep in the ocean.

It can swim to depths of 1280m but the great white is also known to venture into shallower waters and even into the surf. This species is responsible for most unprovoked attacks on humans. However, great whites do not actively hunt for humans and will generally ignore them. A human could easily be mistaken for a seal, or may simply get in the way of an attack on real prey.

Massive megalodon

THE LARGEST SHARK known to have existed (between 16–1.6 million years ago) is the megalodon, a close relative of the great white. Fossils suggests it grew to a length of 16m, had jaws more than 2m wide and teeth up to 21cm long. It's likely that a large megalodon weighed more than 25 tonnes. In comparison, the great white grows to 6m, weighs up to 2.2 tonnes and has 5cm-long teeth.

THAT'S AMAZING

The jaws of a great white shark can deliver a bone-crunching bite that is more than 20 times stronger than a human's. They have the largest teeth of any living shark.

Ground sharks

With more than 250 species worldwide, ground sharks make up the largest order of sharks. Despite this large species variety, these round-bodied sharks are generally similar looking and identification between species can be difficult.

❶ Gummy shark

Mustelus antarcticus

Length: 1.85m Status: Least concern

Australians may be more familiar with this species by the name under which its meat is marketed – flake. In the mid-1980s the species was declared as overfished but regulations have returned populations to a sustainable level.

❷ Bull shark

Carcharhinus leucas

Length: 3.4m Status: Near threatened

This aggressive species has powerful jaws and eats almost anything: sharks, dolphins, rays, fish, turtles, birds and molluscs. It can swim up into freshwater river systems and has been known to take cattle, dogs and people. It's widespread in tropical and warm-temperate seas.

❸ Blue shark

Prionace glauca

Length: 3.83m Status: Near threatened

Named for its distinct indigo-blue upper body that fades to a white underside, the blue shark is a wide-ranging species that prefers cooler ocean temperatures of 12–20°C. Females give birth to live young and have 30–40 pups, although litters of up to 135 have been recorded.

❹ Grey reef shark

Carcharhinus amblyrhynchos

Length: 2.25m (rarely past 1.8m) Status: Near threatened

The inquisitive grey reef shark is one of the most common sharks found in coral reefs. It is known for a unique display – when threatened the shark will wag its head and tail, arch its back, lower its pectoral fins and sometimes swim in a spiral.

❼

Blacktip reef shark
Carcharhinus melanopterus
Length: 1.8m **Status:** Near threatened

Blacktip reef sharks are known to occasionally bite people's feet and legs in the shallows, but are not considered dangerous. Although fairly common, this shark is fished for its meat, fins and oil. However, the species can take up to 16 months to reproduce, so it is vulnerable to overfishing. The species is found in northern Australian waters from Shark Bay, Western Australia to Brisbane, Queensland.

❺

Tiger shark
Galeocerdo cuvier
Length: 6m **Status:** Near threatened

A true scavenger, the tiger shark eats turtles, seals, whales, jellyfish and stingrays, and has also been known to eat livestock, humans and even indigestible objects. It's found in northern waters of various types, from reefs to the open ocean and can be indentified by dark stripes found mainly on juveniles.

❻

Oceanic whitetip shark
Carcharhinus longimanus
Length: 3m **Status:** Vulnerable

This shark prefers warm, deep waters and is found around Australia's north from New South Wales to Western Australia. It eats everything from fish and squid to whales, sea birds and turtles. Some cases of open-ocean attacks on humans are attributed to it.

❽

Bronze whaler
Carcharhinus brachyurus
Length: 2.95m **Status:** Near threatened

This species often occurs in pairs and is a potential danger to humans. It feeds on fish, rays, squid and the occasional sea snake throughout its range in southern Australia from Jurien Bay, Western Australia, to Coffs Harbour, New South Wales.

Farm feed

ALTHOUGH THE meat of the hammerhead is not generally eaten, their fins are highly prized due to a demand for shark fin soup in Asia. There is also a market for hammerhead skin leather, liver oil for vitamins and their carcasses as fishmeal, which is commonly used to feed farm animals.

THAT'S AMAZING

The eyes of this shark are placed on the outer edges of the hammer. This allows it a vertical 360-degree view, which means it is able to see both above and below quite easily. Both eyes are angled slightly forward to reduce the shark's blind spots above and below the head.

Great hammerhead

Sphyrna mokarran

Length: 6m (rarely past 4.5m)

**Status:
Endangered**

EASILY IDENTIFIED by its broad, flat head and tall first dorsal fin, this shark is found in northern Australian waters. While a few attacks on humans have been recorded, the species is not considered to be aggressive. Concerns about the danger of great hammerhead sharks possibly reflect its haunting appearance rather than its actual documented behaviour. In fact, this hammerhead species has a relatively small jaw and small teeth, perfect for eating stingrays, fish and other sea life.

Carpet sharks

From weird and wonderful wobbegongs to the world's largest shark species, the whale shark, carpet sharks are a diverse group. Although differences in appearance can be significant, all species in this order have two dorsal fins and a short mouth that doesn't go past its eyes.

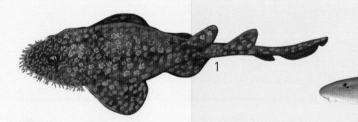

1

2

3

❶ Tasselled wobbegong

Eucrossorhinus dasypogon

Length: 1.25m **Status:** Least concern

With patterned body that blends in with the reef floor, this bottom-dweller inhabits tropical waters from Port Hedland, Western Australia to Bundaberg, Queensland, as well as Indonesia and New Guinea. It feeds on fish and invertebrates, and can bite if provoked or disturbed.

❷ Grey carpetshark

Chiloscyllium punctatum

Length: 1.44m **Status:** Near threatened

Also known as the brownbanded bamboo shark, this species occurs throughout the tropical and subtropical Indo-west Pacific Ocean. Adults are a brown or grey colour while juveniles are pale with about 10 dark, wavy bands.

❸ Blind shark

Brachaelurus waddi

Length: 1.2m **Status:** Least concern

Named for its habit of closing its eyes when caught by fishermen, this shark lives in shallow coastal waters along eastern Australia, between Mooloolaba, Queensland, and Jervis Bay, New South Wales. This relatively common and hardy species is not targeted by commerical fishermen or anglers, and is only occassionally a victim of bycatch. It feeds on reef invertebrates and small fish.

4

Tawny nurse shark
Nebrius ferrugineus
Length: 3.2m **Status:** Vulnerable

These sharks tend to spend their days resting in groups, in caves and rocky crevices. They become active at night, feeding on octopus, crustaceans, sea urchins and small fish, using a powerful sucking action to slurp up their prey.

4

5

5

Zebra shark
Stegostoma fasciatum
Length: 3.5m **Status:** Endangered

This shark prefers warm, deep waters and is found around Australia's north from New South Wales to the coast of Western Australia. It eats everything from small fish to molluscs, crabs and shrimps. The defining feature of the zebra shark is the dark brown spots that cover its body.

6

6

Spotted wobbegong
Orectolobus maculatus
Length: 3m **Status:** Least concern

During the day this species is often found resting in shallow water on reefs or sand, in caves or under piers. Active at night, the spotted wobbegong feeds on reef fish, octopus, crabs and the occasional rock lobster.

Whale shark

Rhincodon typus

**Status:
Endangered**

Length: 12m+

THE WORLD'S largest living fish, this gentle giant is often found near the surface, where snorkellers can swim alongside it. Highly migratory and found in tropical and subtropical oceans worldwide, it appears alone or in large groups.

The whale shark feeds by sucking huge amounts of water in through its massive mouth and out through its gills. As the water passes through, plankton, krill and little fish get caught in special filters at the back of the shark's mouth.

Each whale shark has a unique pattern of spots on its skin. No two whale sharks are identical. This has helped scientists to identify, count and keep track of whale sharks around the world, which helps us to protect this amazing species.

THAT'S AMAZING

Nobody has ever seen whale sharks mate or give birth and no one really knows how many babies a whale shark can give birth to.

Hornsharks

Also known as bullhead sharks, these distinctive-looking sharks are recognisable by their blunt head, small mouth and a prominent crest above each eye. They live on or near the ocean floor in tropical and warm temperate seas.

1

❶ Zebra hornshark

Heterodontus zebra

Length: 1.25m Status: Least concern

Aptly named, the zebra hornshark has a striking pattern of dark, narrow bands on a pale background. It is found in depths of up to 50m in the Western Pacific, from Japan through to the Philippines and Indonesia. In its Australian range on the continental shelf of northern Western Australia, however, the shark generally occurs between 150–200m.

❷ Crested hornshark

Heterodontus galeatus

Length: 1.5m Status: Least concern

The stocky body of this hornshark is mostly yellowish-brown with broad, dark bars on its cheek and below the first dorsal fin. Dark patches also appear along its upper body. The species exists in a relatively small range, from Cape Moreton in southern Queensland to Batemans Bay in New South Wales, and is also possibly found in Cape York. In July and August, females lay 10–16 spiral egg cases with long tendrils that attach to seaweed or sponges. The eggs hatch after about eight months and the young are about 22cm long at birth.

2

Port Jackson shark

Heterodontus portusjacksoni

Status: Least concern

Length: 1.65m (normally much smaller)

WITH A pig-like snout, ridges above the eyes and a harness-like pattern across the shoulder, this is an easily identifiable shark. Frequently seen by divers in rocky gullies and caves from south from the Queensland/ New South Wales border to the Houtman Abrolhos, Western Australia, including Tasmania – it feeds at night on starfish, sea urchins, sea cucumbers and molluscs. It poses no threat to humans unless provoked. Though not deadly, its bite can be painful.

Most sharks give birth to live young but the Port Jackson shark lays eggs in soft, conical egg cases, which the mother will pick up with her mouth and screw into a crevice to prevent them from washing away. After 10–12 months, a baby shark will emerge.

Frilled and cow sharks

Often called 'living fossils', frilled and cow shark species are considered the most primitive sharks because their skeletons more closely resemble ancient, extinct sharks than living species. Cow sharks are easily distinguishable by the presence of six or seven gills, compared to more evolved species that have five. Just two species of frilled shark survive today, and only one of these is present in Australian waters.

❶ Sharpnose sevengill shark

Heptranchias perlo

Length: 1.39m Status: Near threatened

Populations of the sharpnose sevengill shark are not large but the species is found in tropical and temperate regions of all oceans except for the north-eastern Pacific. Despite its small size, this shark is a top predator within the ecosystems it inhabits. Off Australia, it feeds primarily on bony fish, squid and octopus. If captured it is aggressive and will try to bite but cannot do much harm to humans.

❷ Bluntnose sixgill shark

Hexanchus griseus

Length: 4.8m Status: Near Threatened

Also known as the cow shark, this large, heavy-bodied shark spends most of its time near the ocean floor at depths of up to 2500m. It is a nocturnal feeder known to travel to shallower depths but will only come to the surface at night when it is dark. It has a broad, flat head, rounded snout and large green eyes.

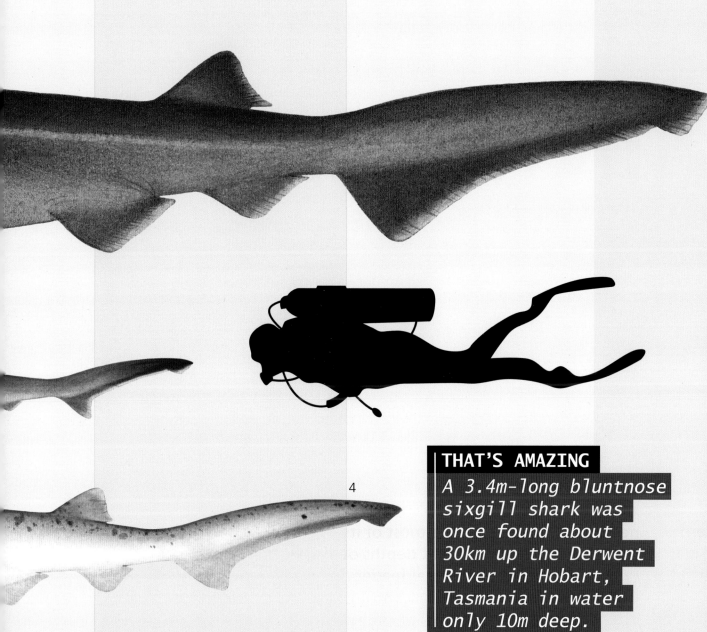

❸

Bigeyed sixgill shark
Hexanchus nakamurai
Length: 1.8m Status: Data deficient

Little is known about the behaviour of this slender shark, which is patchily distributed throughout tropical and warm temperate waters of the Indian and western Pacific oceans. Its large eyes glow a fluorescent green and it has five rows of large, comb-like teeth on each side of its lower jaw.

❹

Broadnose sevengill shark
Notorynchus cepedianus
Length: 3m Status: Data deficient

This shark occurs in most temperate seas, found in southern Australia from the central coast of New South Wales around to south-western Western Australia. Larger individuals have been known to hunt seals and cetaceans, using stealth to sneak up on their prey. Pack-hunting behaviour has also been observed. Female broadnose sevengill sharks give birth to large litters of up to 82 pups.

THAT'S AMAZING
A 3.4m-long bluntnose sixgill shark was once found about 30km up the Derwent River in Hobart, Tasmania in water only 10m deep.

4

In focus:

Frilled shark
Chlamydoselachus anguineus

Status: Least concern

Length: 1.96m

THE FRILLED shark is found across a wide range but it is pretty patchily distributed – throughout the eastern, central and western Pacific Ocean, along with the North Atlantic. This shark has an eel-like body, a short snout, a long mouth and six gills.

It looks extremely similar to a species found in southern Africa. The two can only be distinguished by looking at features inside the body.

Little is known about the biology and behaviour of this deep-ocean dweller. With a large mouth and jaws that can open very wide, it is capable of eating relatively large prey. It is believed that females undergo a long gestation period of up to 3.5 years, the longest pregnancy of any vertebrate animal, before giving birth to an average of 6 pups at once.

The frilled shark's mouth is just as terrifying as the jaw of a great white: It's lined with 25 rows of backward-facing, trident-shaped teeth – 300 teeth in all!

Angelsharks

Of the 18 species of angelshark, four are found in Australian waters. These sharks resemble rays because of the broad, wing-like pectoral fins that extend from their body. Their eyes are located on top of their heads and their mouth on the end of their snouts. Angelsharks are mostly found relatively close to the shore, where they bury themselves in the sand and wait for prey such as fish, crustaceans and molluscs.

1

❶

Eastern angelshark

Squatina albipunctata

Length: 1.3m **Status:** Vulnerable

Featuring spines near the eyes and fringed barbels on the snout, the eastern angelshark is covered in fine, white spots. This species is only found off eastern Australia between Cairns, Queensland and Lakes Entrance, Victoria, at depths of 35–415m. It often ends up as bycatch due to its habit of lying in the sand on the sea floor.

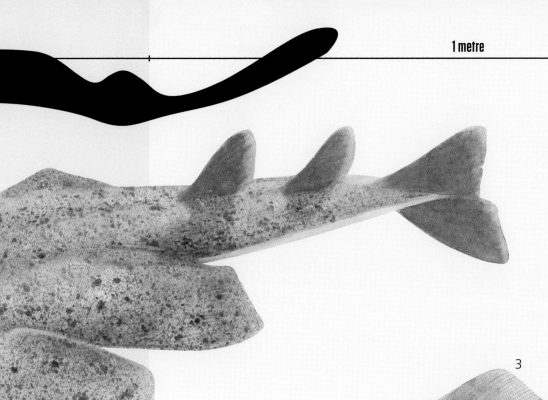

❸

Western angelshark
Squatina pseudocellata

Length: 1.14m   Status: Least concern

Found exclusively off the West Australian coast, from Cape Leveque to Shark Bay, the western angelshark inhabits the tropical outer continental shelf. Its brown or greyish upper body is covered with widely spaced bluish spots and darker brown patches. In the past this species has been confused with the similar-looking ornate angelshark.

3

2

❷

Ornate angelshark
Squatina tergocellata

Length: 1.4m   Status: Least concern

The ornate angelshark has a unique pattern. The spots on its surface, known as ocelli, look like cells dividing. These sharks feed mainly on squid, particularly the red arrow squid, and small fish known as leatherjackets. The ornate angelshark only lives in southern Australian waters.

The whiskery barbels near the angelshark's mouth are for sensing nearby prey – they are covered in taste buds!

Australian angelshark

Squatina australis

Status:
Least
concern

Length: 1.52m

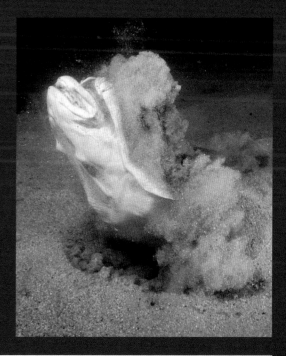

A COMMON VISITOR to beaches along the southern coast from Lancelin, Western Australia, to Sydney, New South Wales, the Australian angelshark inhabits depths of 150–310m. It can be recognised by its flat body with large, wing-like fins that don't fully attach to its head.

This nocturnal shark uses long, sharp teeth to feed on its diet of fish, squid and crustaceans, which it ambushes from a place of camouflage on the sandy ocean floor. During the day it lies buried in sand or mud, often in seagrass beds or near rocky reefs, waiting for its prey.

Sawsharks

It's easy to see how the sawsharks got their name – a long, blade-like snout extends from their bodies, armed with sharp, horizontal, tooth-like denticles. These 'saws' are used to slash and disable prey. When they are growing inside the mother, the denticles of sawshark pups are inverted into the mouth to prevent any harm to her. Three species of sawshark occur nowhere else but Australian waters.

❷ Tropical sawshark

Pristiophorus delicatus

Length: 0.84m Status: Least concern

Not a great deal is known about this small, yellowish-brown shark, which was first officially described in 2008. It is known to occur off tropical north-eastern Queensland, from Rockhampton to Cairns. Females grow to at least 84cm but no adult males have been found, only adolescents at 63cm.

❶ Southern sawshark

Pristiophorus nudipinnis

Lenth: 1.24m Status: Least concern

Compared with some related sharks, the snout of the southern sawshark is quite small. It uses the barbels located along its snout to find food on the ocean floor before stunning prey with vigorous movement of the snout.

THAT'S AMAZING
There are eight species of sawsharks in the world. Their saws have special sense organs that pick up the electric fields of prey, even when buried in sand.

Common sawshark

Pristiophorus cirratus

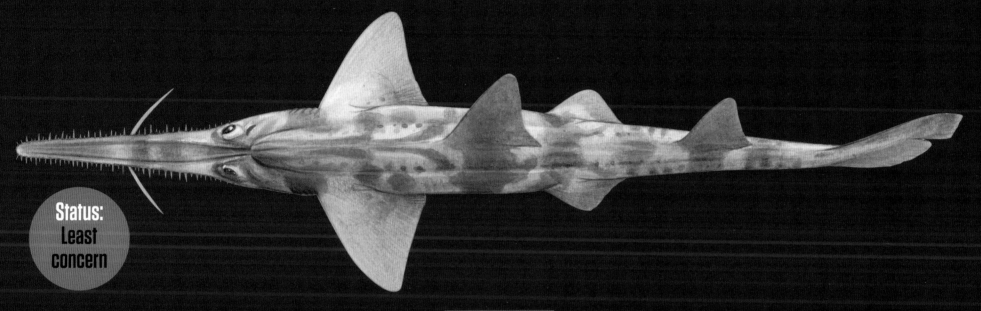

**Status:
Least
concern**

Length: 1.49m

THE LARGEST and most widely distributed of Australian sawsharks, the common sawshark is covered in a pattern of brownish blotches and dark bands on a sandy body. The markings of common sawsharks in eastern Australia tend to be much less pronounced, however.

Once every two years, after a 12-month gestation period, female sharks give birth to a litter of 6–9 pups. Endemic to Australia, the common sawshark uses its whisker-like barbels and snout to detect prey on the ocean floor, grabbing it with its large mouth and rows of small teeth.

Dogfish and bramble sharks

Generally bottom-dwelling sharks, this group includes dogfish and sleeper sharks, as well as species from the only shark families known to glow – the lantern and kitefin sharks. Bramble sharks were once placed in the same order but have since been described as a different group, based on dental, skeletal and muscular differences. They are named for the thorn-like denticles that grow over their body.

❶ Spiny dogfish

Squalus acanthias

Length: 2m Status: Vulnerable

This slender-bodied shark is named for the single spine located in front of both fins on its upper body. Despite its natural abundance, populations have been significantly depleted due to demand for its meat.

❷ Southern sleeper shark

Somniosus antarcticus

Length: 4.56m Status: Least concern

This deep-ocean dwelling shark reaches at least 4.5m in length and reportedly more than 6m. Despite a sluggish appearance, this shark is believed to target giant squid – but it is not known whether the shark catches the squid while it is still alive.

❸ Bramble Shark

Echinorhinus brucus

Length: 2.6m Status: Data deficient

The bristles that cover the bramble shark are large, irregularly and patchily distributed, and sometimes fused together. This purplish-brown shark is a sluggish, primarily deep-water species that is usually covered in foul-smelling mucus when caught.

5

Southern lanternshark

Etmopterus baxteri

Length: 0.9m **Status:** Least concern

This species is a small shark, with a short snout and large eyes. Also known as the New Zealand lanternshark, it is found in deep waters off southern New South Wales, Victoria and Tasmania, as well as New Zealand and southern Africa.

6

Kitefin shark

Dalatias licha

Length: 1.6m **Status:** Vulnerable

This powerful predator uses its large teeth and strong bite to capture a range of prey, including bony fish, rays, crustaceans, cephalopods and even other sharks.

7

Cookiecutter shark

Isistius brasiliensis

Length: 0.5m **Status:** Least concern

The cookiecutter shark has an impressive feeding technique. Opening its mouth wide, it attaches to the surface of its prey, then swivels its body so that its teeth cut out a round, cookie-shaped piece of flesh.

4

Prickly shark

Echinorhinus cookei

Length: 4m **Status:** Data Deficient

The smaller and more numerous denticles of the prickly shark are evenly distributed over its body and do not fuse together. This species has been observed hovering almost motionless just above the sea floor but the reason behind this behaviour is a mystery.

8

Prickly dogfish

Oxynotus bruniensis

Length: 0.75m **Status:** Near threatened

The unique-looking prickly dogfish is characterised by rough skin, sail-like fins and a hump on its back. It is thought to swim slowly above the ocean floor and hunt for small invertebrates and fish.

9

Harrisson's dogfish

Centrophorus harrissoni

Length: 1.14m **Status:** Endangered

The rare and endangered Harrisson's dogfish, or dumb gulper shark, is found only along the eastern Australian coast and in isolated occurrences north and west of New Zealand. Its teeth are much larger on the lower jaw than the upper, and the upper teeth stand much more upright in males.

Pygmy shark

Euprotomicrus bispinatus

Length: 0.27m

THAT'S AMAZING

The pygmy shark is the second smallest shark in the world. The smallest is the dwarf lanternshark which is found off South America and grows to a length of 20cm.

BLACK WITH pale fins, a luminescent belly and an underslung jaw, the pygmy shark measures less than 30cm when fully grown and is harmless to humans. An open-ocean dweller, it spends daylight hours in deep water (to depths of 1520m) and migrates after sunset to the surface in pursuit of bony fish, cephalopods and crustaceans.

In Australian waters, it's found in tropical and warm-temperate seas from Perth to Rowley Shoals, west of Broome, Western Australia.

This deep-ocean dweller is bioluminescent – meaning its body glows – an adaptation believed to be a camouflage technique in which the shark blends in with light filtering down from the surface.

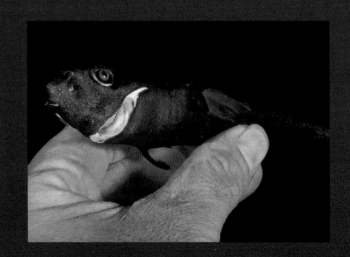

ICONOGRAPHIA ZOOLOGICA/WIKIMEDIA